HOW TO BE A ROCKET SCIENTIST

10 powerful tips to enter the aerospace field and launch the career of your dreams

BRETT HOFFSTADT

"What did Archimedes say?
'Give me a lever long enough and I'll move the world.'
The lever is knowledge and the knowledge is yours
when you've read through Brett's book."

Nelson Louis Olivo, Chairman
Young Astronauts Program, United States of America

"It draws you in, couldn't stop reading it. Precise and up to date
advice with great resources!"

Jackelynne Silva-Martinez,
Mechanical and Aerospace Engineer,
NASA Jet Propulsion Laboratory

*"Once you have tasted flight
you will walk the earth with
your eyes turned skywards,
for there you have been
and there you long to return."*

Leonardo da Vinci

DEDICATION

To all of those who have been possessed with the urge

to fly, to make things fly, or to take humanity

into the air and space above.

We came from the stars,

and to the stars we shall return.

CONTENTS

PREFACE TO THE 2021 EDITION

This book was originally published in 2014. It's been a remarkable seven years in the world of "rocket science."

At the end of 2014, SpaceX had made four attempts toward the vision of a reusable orbital rocket. None of these attempts to land and recover a first stage booster had succeeded.

As of September 2021, they have achieved it 92 times! Their Falcon 9 rocket has created a new paradigm (and a transformational business case) for access to space.

We've seen Virgin Galactic and Blue Origin both reach space with their billionaire founders (literally and figuratively). Space tourism is now a dedicated sector within the space industry.

Besides the International Space Station (ISS), we also now have a fully operational Chinese space station orbiting Earth. Three taikonauts spent 90 days onboard there in 2021.

We are in a global space renaissance, and we are just getting started. According to space.com, there are at least nine companies working on lunar landers or rovers (or both). This time, it will be to stay. There are at least ten companies pursuing asteroid mining.

Despite this new world, as I study the ten tips to be a "rocket scientist" that I advised in 2014, they are still extremely relevant and effective. The feedback and stories that I have received from my readers prove this out.

These 10 tips combine to give you two powerful messages. First, you don't need to wait for anyone's permission to begin your education or your career in "rocket science." You can (and you must!) pick yourself. Secondly, the world of "rocket science" is large and diverse enough that, whatever your strengths and passions and background, there can be a way for you to chart your own unique path into this "space."

Tim Dodd earned his living in Iowa as a motorcycle mechanic and a wedding photographer, very far from any aerospace hub. He applied his photography skills to space exploration, creating portfolios on these themes. This led to a freelance job as a photographer for space missions. Then he created a YouTube channel to be a communicator for space science. His channel, *Everyday Astronaut*, has reached over one million subscribers in 2021. (Now it brings serious revenue for him.)

In September 2021, he scored a major career win as the official livestream host (i.e. temporary public relations director) for Firefly Aerospace as they made their first launch of their Alpha rocket.

If you sought traditional education and career advice to enter aerospace or aviation, would you ever find, "Be a wedding photographer. Then start a YouTube channel. Then you'll get hired by an aerospace company to represent them during their most important event in the history of their company."

Of course not!! And yet, that's what happened to Tim Todd. And yes, he can say he is a "rocket scientist" in today's world.

Think about him as you read through these 10 tips. You should see how they contributed to his unique niche in the aerospace world.

More importantly, I urge you to think about how you can apply these 10 tips for your own success, happiness, and impact.

The examples in this book may seem a little dated, but they are still timeless. If you want examples that are more recent or more abundant, I encourage you to go to the blog page at my website: www.howtobearocketscientist.com. As of September 2021, there are 177 blog articles with other examples, stories, and resources that reinforce the 10 tips you are about to learn.

(By the way, through my own efforts to help the organization **Teachers In Space**, which had a cubesat payload on the Alpha rocket, I was fortunate to be included in the launch observation group for the Alpha launch. You can get my first-hand account of that amazing experience on my blog too.)

There's one more thing about rocket science that hasn't changed which I need to emphasize to you. Rocket science still ain't easy! Sorry to disappoint you, perhaps. Charting your own career path into this field isn't going to come easy either. It will take hard work. Persistence. And courage. But the belief that you can create a path will hopefully motivate and inspire you to work through your fears and struggles.

The rewards are out there. So are the opportunities. There are millions of aerospace and aviation professionals who, like me, have a passionate desire to see more people enter our community. We know there are good people out there. That means you! I hope this book is a uniquely valuable tool for you to transform your dream into reality.

Brett Hoffstadt
September 25th, 2021

ACKNOWLEDGEMENTS

While it is certainly true that one person can make a difference, it is never without the efforts and benefits from many other people who blazed the path beforehand and who contributed or influenced the effort in some way. That is certainly true for building rockets, aircraft, and spacecraft! It's also true for any book. To everyone I've had the privilege of studying with, working with, and learning from, I offer my profound appreciation and gratitude. I also salute the thousands of others who played roles—large and small—to lift humankind to greater heights above the Earth. Even the smallest part had a valuable purpose.

My deepest thanks to my parents, Fred and Nancy Hoffstadt, who lovingly encouraged and supported me with flying toys and projects of all kinds as I grew up. Profound thanks to my wife Rita. She's been a very rare and treasured person to spend many years growing together with and sharing the journey. And thanks to Maya and Max for being the most fun and rewarding "people projects" I've ever had (besides myself). A final special thanks to everyone who has contributed to the production, editing, feedback, and conversations surrounding this book. It all continues after the initial printing—as it is meant to do.

Brett Hoffstadt
San Antonio, Texas
December 2014

WHY REACH FOR THE SKY?

There simply is something timeless, profound, and irresistible about looking up into the sky, whether it's on a sunny breezy day or on a night with the moon and stars in plain sight. What could be more awesome than knowing how to get up there? Maybe doing it yourself someday?

Aerospace engineers and scientists, and many other people who work in aerospace fields, make a difference that very few others can. Whether you are still a student or a workforce veteran contemplating a long-sought career change, this book will give you first-hand, "inside," and straight-shooting information on what it takes to be successful in a career that starts with those skyward dreams.

Let's get some definitions clear at the start. The aerospace world involves a whole lot more than just rockets. It has a lot more than science too, as you'll soon read. But collectively, this industry is often called "rocket science," with the people who work in it half-jokingly called "rocket scientists." Let's not debate or resist the name—let's embrace it! Whether it is aeronautics, aviation, or space flight—the skills and knowledge used in these domains are all part of the aerospace education system, profession, and marketplace. Anyone who has a valuable role in these sectors can be called an aerospace professional (and hopefully they act like one). Since "rocket science" is a commonly-used synonym or substitute for aerospace, this book uses "rocket scientist" to also mean "aerospace professional." Keep this larger context in mind as you discover and explore the wide variety of careers that can be made here.

Some might claim that the best or biggest days of rocket science are behind us. Not true at all!

In some important and undeniable ways, the *first* golden age of space exploration has faded into the past. USA and Russian space programs and major aerospace companies are still around, but they have some serious competition on the scene. We now have the "new space" or Space 2.0 industry in full force. This new era is typified by private companies such as Space-X, Virgin Galactic, Bigelow Aerospace, and many others. They are racing ahead to find profitable solutions for space tourism, asteroid mining, and space-based power.

One major subset of the aerospace world is aeronautics. Aeronautics pertains to everything man-made that flies within our atmosphere. Big companies and agencies still exist for commercial transports, business jets, helicopters, and military aircraft. But we also have light aircraft companies, constantly innovating suppliers for systems and components, and the relatively young market of unmanned aircraft. Unmanned Air Vehicles (UAVs) are poised to become a revolutionary force in aerospace much like the jet engine revolutionized aviation after piston-powered propeller-driven aircraft.

We should be realistic along with our optimism. We can't expect all of these companies to succeed. But here are some other facts that you need to know. A survey of Aerospace & Defense (A&D) companies in late 2013 by people at *Aviation Week* (the go-to weekly trade magazine) found that 9.6% of all employees were eligible to retire. That was 62,000 people! By 2017, predictions show that 18.5% of the A&D workforce will be eligible to retire. Imagine 1 out of 5 people vanishing from an industry! Who's going to do all of that work?!

When I listen to and interact with managers and executives in aerospace (often at conferences, which you'll read more about in Tip #3), I can tell you that they are very aware and very concerned about this unavoidable demographic cliff. The slow economy is motivating many of the Baby Boomers to work longer than they prefer. But conditions always change, and nobody can stick around forever, can they?

If you take the long-range view and want a career in aerospace, you have a solid job market opportunity. Plus, compared to other engineering fields and specialties, aerospace is consistently near the top for average salaries. And of course, rocket scientists have the "cool" factor.

There is a constant stream of articles and stories about a shortage of science, technology, engineering, and math (STEM) workers. It is sometimes hard to reconcile that with the hard facts about unemployment and underemployment today. But of all the STEM fields, aerospace has a lot going for it. First, it is less vulnerable to foreign competition (outsourcing) because of the sensitive and proprietary nature of the work. In the United States, major defense companies like Lockheed Martin, Boeing, Northrop Grumman, and Raytheon MUST hire US citizens or green card holders for their military space and aircraft programs. Other countries are similarly possessive about their own aerospace technologies and industries if you are a citizen there.

In summary, why pursue a job or career in rocket science and aerospace? Yes, it is frickin' cool. Yes, the money is good. Yes, your parents and family will be proud. And yes, you'll have a great pick-up line or ice-breaker at any party! Or you'll be hanging out with other rocket scientist pals. How fun is it to know and say that you are overcoming gravity—one of the most awesome forces in the universe?

The simple fact is, we need you and more people like you. It doesn't have to be an impossible dream. This book will show you how to make it happen—if you are willing to start and persist.

In fact, if you are someone not typically associated with engineering and STEM, you are even more desirable. That's because a diverse workforce is proven to bring benefits to the bottom line. Innovation and creativity are desperately needed in aerospace—just like other industries (along with quality and controls). A Boeing colleague of mine once gave a beautiful quote: "The most interesting problems are solved by the most interesting people." (I wish I could give credit for that but I heard it second-hand...I'm still trying to track that person down.)

Aerospace and aviation are full of interesting problems!! So the more interesting or unique you are, the better! (But there are still some social and people-skill topics to discuss, which we will do in Tips 3, 9, and 10.)

You can see that there are lots of reasons to go for it. And we need you! There are still many fascinating problems to solve, innovations to create, and far away places to explore. So let's get moving and get started!

P.S. If you don't find the answer to your question or you want more details, I invite you to contact me. My contact information is at the end of the book. I just ask that you read the entire book first before reaching out to me. (And thanks again for purchasing it!)

P.P.S. Since this is a paperback edition, it will take you a little more time and initiative than using the eBook to find the resources on the web that I mention. I wouldn't provide them if I didn't believe it was worth it to you. Achieving your dreams will take initiative and work on your part. I hope you'll pursue them with an attitude of eagerness and determination. *Now* let's get moving with how to be a rocket scientist!

DISCLAIMER: Works, statements, and products from other people mentioned in this book do not signify official endorsements of or agreement with the entire work. Furthermore, no endorsement or agreement with their statements, beliefs, or positions outside of their work is made or implied. Likewise, these recommendations have, in most cases, been made without their knowledge or approval. You should not assume they have authorized or sanctioned my recommendations. (Rocket scientists need to care about legal matters too.)

TIP #1:

REFLECT ON YOUR PASSIONS

Let's be honest with ourselves—and with each other. Rocket science isn't always easy! Neither is having a career in this field. There are forces beyond your control such as your genetic identity, the macro economy, family needs, etc. On the other hand, there is a tremendous amount that is IN your control: your thoughts, actions, and attitudes! Let's get into the fuel that will keep you moving and into the guidance, navigation, and control (GNC) signals that will keep you moving in the right direction.

It is going to take courage, initiative, determination, persistence, and grit. (You are surprised about courage? Check out the book by Allan McDonald in Tip #4.) A healthy sense of humor will be very helpful too.

There is almost certainly an element of luck involved. But luck has to be both identified and then acted upon to make it work for you! As Ben Franklin said, "...the harder I work the more I have of it [luck]."

Now to the fuel that will propel you into or through or around all of these situations. Ferdinand Foch, a French General from World War I, said, "The most powerful weapon on earth is the human soul on fire." It's time to activate and engage your weapon—and only you can do it! Identify, assess, and embrace your passions and you will find the fuel to set your soul on fire. It is with this passion and internally-driven enthusiasm that you will find the willpower to persist and the resourcefulness to jump on "lucky" opportunities when they appear.

Not sure how to find your passion or identify it? Then use these tips:

- Think about or be conscious of when you feel "in the zone." When does time fly by?

- If you are a morning person, what would make you stay up late?

- If you are a night owl, what would make you jump out of bed early in the morning?

- What kind of work, activity, or job would you enjoy doing so much that you would laugh with amusement if (or *when*) you got paid for it?

There are no wrong answers as you build your list. When you assess this list, you want to apply a filter for ethics and legality. For now, get in touch with your passions, your loves, and whatever sets your soul on fire. Then—very important— *write them down!*

Bonus tip and next steps: once you have identified one or more passions related to rocket science or aerospace, re-assess how you spend your leisure and recreational time. There are PLENTY of fun things to do here: radio-controlled airplanes, autonomous drones, model rockets, cubesats (small satellites), local clubs…put your passion into action! Learn more and *be more* by **doing more!**

Additional Resources:

- Knowing your passions is one early step in the best-selling and annually-updated career book, *What Color is Your Parachute?* by Richard Bolles.

- Why is passion so important? Watch the powerful TED Talk from Simon Sinek, *Start With Why.*

- A classic reference book (written at a college-level, very academic style) is *Flow: The Psychology of Optimal Experience* by Mihaly Csikszentmihalyi.

- A modern classic (also great for a wider audience like teachers and parents) is *The Element: How Finding Your Passion Changes Everything* by Sir Ken Robinson.

- Joseph Campbell coined the phrase "Follow Your Bliss." Learn more at about him from the Joseph Campbell Foundation at www.jcf.org.

TIP #2:

EXPAND YOUR CONCEPT OF
ROCKET SCIENCE

It takes many people to make and launch rockets. And airplanes. And air traffic control systems. Many engineers, scientists, and technicians. But it also requires many financial analysts, managers, computer/IT staff, graphic designers, marketing experts, attorneys, human resource professionals, salespeople—plus many more jobs besides these!

Whatever skills or expertise you have and enjoy, or want to pursue, you should be able to find someone in the aerospace world who has already done it. (But read Tip #9 first!) Or you should be able to find this special skill or talent of yours exhibited somewhere in the aerospace field.

Food scientists work at NASA to develop foods that can last for years in space—and still hold their nutrition and flavor.

Interior designers work on VIP jets—and earn a major salary with trips to exotic places so they can visit with their VIP customers.

Marketing and sales people work in every aerospace sector to sell products and services.

Heating, Ventilation, & Air Conditioning (HVAC) experts design heating and cooling systems for things that fly. (When you earn a college degree you get to be called a Thermal Analysis & Design Engineer.)

All of the buildings where people work have all of the support and maintenance staff that comes with every other building in the world—plus other fun specializations like zero dust environments, electromagnetic shielding, and restricted access controls (the serious security stuff).

Biologists help design the sensors that dig into Martian rocks, analyze the results, and look for signs of life beyond Earth.

All of these people and skills are needed in the world of aerospace and rocket science. If you are one of them, you get to say you work in rocket science! Compared to everyone else, that makes you a rocket scientist.

Additional Resources:

- A list of the major disciplines in aerospace can be found on Wikipedia (www.wikipedia.org). Search for "aerospace engineering" and you'll find "Elements" as one section on the page. And there are many more! But you just might be better off creating your own unique niche anyway—which we'll talk about in Tip #8.

- The Career Cornerstone website page for aerospace has interviews, earning data, a podcast, online links, and much more! This one website can keep you busy for a very long time. Go to www.careercornerstone.org, then go into Engineering to find Aerospace specialty.

TIP #3:

ASSOCIATE WITH PEOPLE IN THE FIELD

For any career path, the more you hang around, interact with, and are recognized within the field, the more likely and natural a transition into that field will be. The same is true with rocket science.

Not only will you meet people, you'll get to learn the language and lingo of the industry. Jargon and acronyms—every industry and profession has its own. Aerospace is no different—especially the acronyms! We *love* acronyms in aerospace. I wanted to learn another language, but that part of my brain was already full with acronyms that I used at work. Oh well....

You also get informed about current events, trends, issues, opportunities, market forces...if you are passionate about doing this for a living, wouldn't you want to know what is happening in the profession and industry?

How can you do that if you aren't working in the field yet? There are lots of ways!

In person, there are professional societies with local chapters or sections in almost every major city. They hold events such as dinner meetings, company tours, education outreach events, and social gatherings. The largest is the American Institute of Aeronautics and Astronautics (AIAA). Go to their website, find the nearest local Section, and get involved! Get out of your house or apartment!

Depending on your more specific interests or things that happen in your area, there can be other groups, too. Get ready for some acronyms!

There are groups that are industry-specific or sector-targeted or application-focused. At the highest level is the Aerospace Industries Association (AIA). The most promising sectors in upcoming decades appear to be unmanned aircraft and commercial spaceflight. The central organizations for each of those are the Association for Unmanned Vehicle Systems International (AUVSI) and the Commercial Spaceflight Federation (CSF). For other interests, you can find the American Helicopter Society (AHS), the Academy of Model Aeronautics (AMA), the Civil Air Patrol (CAP), the European Aviation Club (EAC), the Experimental Aircraft Association (EAA), the National Space Society (NSS), and The Planetary Society (TPS). And that isn't all of them! If you have passion around a particular purpose or market within aerospace, search the web for "association," "society," or "institute" added on to that topic. What seems like a geeky fringe obsession of yours just might be the focus of an entire group of future friends and colleagues.

If you care more about a specific discipline, the major engineering societies usually have an aerospace committee or component to them: SAE (automotive), IEEE (electronics), ASME (mechanical engineers), ASM (metals)...plus many more.

There are also organizations dedicated to particular demographics or ethnicities. For example, there is the Society of Women Engineers (SWE), National Society of Black Engineers (NSBE), and the Society of Hispanic Professional Engineers (SHPE). If talking to someone in one of these groups is a more comfortable way for you to start learning about an engineering career, go for it!

All of these organizations also host large conferences. There are technical sessions, executive panels, exhibit halls with company representatives you can approach, product demonstrations, and more. Once you get involved in one of these professional societies or organizations, you will learn about these opportunities. The networking, casual conversations, and serendipitous occasions (so-called "lucky breaks") are priceless.

If you are under age 18 or 21, most of these organizations have a student membership or an education foundation (or both) that caters to people your age. ("Students" for some groups will also include college students.) Ask for help from an adult family member or teacher or community member if you need it to join or get more information from the group you are interested in.

In print or writing, there are lots of options. If you can afford it, *Aviation Week* is the industry's go-to weekly publication. Yes, it is a weekly magazine. It could be worth a library trip if one of your local libraries subscribes to it. (Or if you have a college or university within target range and they have an engineering school, try that.) Remember that list of professional societies for certain disciplines? Most of those have their own magazines and publishing division that provide great information and insights.

On email, there are many resources here, too. Again, most or all of the large professional societies provide an email news digest to members. Some offer it for nonmembers too. Two that I use and recommend are the Daily Launch from AIAA and the dailyLead from AIA.

In social media, there are groups on LinkedIn, Facebook, Twitter, etc. LinkedIn is the current defacto social media site for professionals (or a person's professional identity). Set up a profile for yourself as a respectable, teachable, honest professional (perhaps at a very early stage in your career), and you will be making strong progress.

If you are already in the workforce or actively seeking employment, LinkedIn is the site you need to have a presence on and use as a tool for your own research.

There could be a lot more said about how to manage and use your social media and online presence. At this point, let's just remind ourselves that *anyone* can do an internet search on you! Think about the digital trail you are leaving behind and the future you want to project.

The big benefit of social media is that you can start associating with people or groups today. Find and follow the Twitter feeds of people or companies you are curious about. Follow people or companies on LinkedIn to get updates from them. Many companies use other social media, too.

However and whatever is most convenient and comfortable to you—just get started! But don't plan to stay comfortable for too long. If you are looking for new opportunities and changes for yourself, by definition that means moving into areas or spaces that aren't familiar to you. The sooner you get started, the sooner it will become a comfortable environment for you.

Additional resources:

- The American Institute of Aeronautics and Astronautics (AIAA) is the world's oldest and premier professional organization for aerospace professionals. (www.aiaa.org)

- Aerospace Industries Association (AIA) is the industry-wide voice for business in the United States. They provide a free daily email digest that is very informative. If you are outside the US, look for similar associations in your country or region. (www.aia-aerospace.org)

- The Institute of Electrical and Electronics Engineers (IEEE) is the world's largest professional society for electrical, computer, and software engineering. This includes avionics which is the term for aviation-based electronics. (www.ieee.org)

- ASM International was founded as the American Society for Metals, but now has a broader geographic reach. (www.asminternational.org)

- For composite materials, two key groups are the American Society of Composites (ASC, www.asc-composites.org) and The Society for the Advancement of Material and Process Engineering. (SAMPE, www.sampe.org)

- The American Society of Mechanical Engineers (ASME) is the premier organization for this discipline. (www.asme.org)

- The premier organization for unmanned vehicles (not only in the air but on land and in the water too) is the Association for Unmanned Vehicle Systems International. (AUVSI, www.auvsi.org)

- The organization dedicated to vertical flight, including helicopters, V/STOL (Vertical, Short Take-Off and Landing), whether pilots of unmanned, is the American Helicopter Society. (AHS, www.vtol.org)

- The go-to organization for commercial or private spaceflight is The Commercial Spaceflight Foundation. (CSF, www.commercialspaceflight.org)

- The Society of Women Engineers (SWE) is online here: www.swe.org.

- The National Society of Black Engineers is online here: www.nsbe.org.

- The Society for Hispanic Professional Engineers is online here: www.national.shpe.org.

See something important missing? Contact me through a channel mentioned at the end of this book so I and other readers can learn about it.

TIP #4:

READ A BOOK ON ROCKET SCIENCE

There are epic stories in the aerospace world. They are as good as the best science fiction novels—but they are science reality! And they also have epic, historic characters.

For instance, did you know that it's possible to eject from an airplane travelling at three times the speed of sound? (And live to tell about it…you'll see in Richard Graham's book.) And this aircraft, the SR-71 Blackbird, travelled so high and so fast that a new navigation system had to be developed that used other stars as reference points! And this was in the 1960s, before we landed on the moon!

There are also some other great non-fiction books that give you rare, practical insights and advice on an aerospace career (besides this one). Here is a short list of some powerful reading. Pick one and get started.

Advice to Rocket Scientists by Jim Longuski

This is required reading, no matter what stage of your career you are in! This book is full of great nuggets of wisdom, written from an aerospace industry veteran and a professor at Purdue University. (He is also one of my favorite professors who I had the privilege of learning from.)

SR-71 Blackbird: Stories, Tales, and Legends by Richard Graham

This book is written by one of the very few SR-71 pilots and instructors. You can read about the once secret activities surrounding one of the greatest aircraft ever designed, built, and flown.

The Seven Secrets of How to Think Like a Rocket Scientist, by Jim Longuski.

Read more great tips from this professor at Purdue University who previously worked on the Cassini space probe at the Jet Propulsion Laboratory (JPL).

Rocket Rangers: Man's Quest to Fly Like Buck Rogers, Volumes 1 and 2, by Nelson Olivo

These two volumes will inspire you no matter what your age, while also providing you with incredible first-hand accounts and full-color historic photos related to solo flight. Volume 1 features jetpacks, rocket belts, and more. Volume 2 has first-hand accounts from astronauts who flew in Earth's orbit—completely disconnected from any spaceship!

Truth, Lies, and O-Rings: Inside the Space Shuttle Challenger Disaster by Allan McDonald & James R Hansen

Do not check your character, integrity and courage at the door! This is a gripping and reflective first-hand account from someone who knew the Challenger space shuttle shouldn't launch in such cold temperatures and tried (unsuccessfully) to delay the launch.

The Dream Machine by Richard Whittle

This is the long and winding story of the V-22 Osprey. It reads like a gripping suspense novel, complete with second-by-second action sequences. It also describes the intriguing and true-life reality of politics and personalities inside major aerospace programs. It's more interesting and appropriate for college-age readers or young professionals.

Engineering Stories by Ken Hardman

How about some short stories? This technical expert and leader in the aerospace industry has created a number of "realistic engineering adventure stories." Many individual stories are available online at http://engineerstories.com. If you subscribe to his blog, you'll get more stories as he writes them.

Are there any other books that you think are essential for a successful launch into an aerospace career? Please let me and others know through my website blog, email, or Twitter (found at the end of the book).

TIP #5:

WATCH A MOVIE ABOUT ROCKET SCIENCE

This may not seem like a serious and worthwhile tip. But think back to the wise words of General Foch. We all need to stay inspired, motivated, and persistent in our journey of life. Movies are powerful experiences to help us remember why we do what we do, and why it's worth pushing on. Movies help light our souls on fire—and keep them lit!

Also, more practically, there are movies that are cult classics for many rocket scientists, flyboy/flygirl geeks, and aerospace nerds. It's smart to have a shared experience through these movies. You'll have a better understanding about what inspires this crowd and what we love to watch.

First, if you want a free and short online video that gives you an inspiring and dramatic overview of the aerospace world, go to the **AIAA website (www.aiaa.org).** Then do a search for the term **"from the sands to the stars."** That should bring you to a 5-minute video with that name. It's a promotional video for AIAA but it's also a great video I've used at career day and classroom events. (If you are very observant, you'll notice something unusual during the video. You'll have to watch it to try and find it.)

Flight of the Phoenix (1965)

This version with Jimmy Stewart is far better than the 2005 version (in my not-so-humble opinion). What you can get out of this movie: 1. It isn't enough to be smart about something-- you have to work well with others or you're going to get stuck in a very bad place. (See Tip #10) 2. Engineers can save the day!

Cosmos (original series in 1980, modern series in 2014)

This isn't a movie, but—thanks to Carl Sagan—it inspired an entire generation of kids to look skyward and pursue careers in STEM. The entire 13-hour series from 1980 is available on YouTube! The modern recreation of this show has been broadcast on Fox and National Geographic with Neil DeGrasse Tyson as the narrator and host.

The Right Stuff (1983)

You'll see amazing and true stories from the glory years of America's launch into the space age. It's an epic movie with historic figures and great narratives.

Apollo 13 (1995)

This is a blockbuster movie which shows the genuine courage, innovation, and teamwork needed to fly into space. And—more importantly—how to return safely.

Contact (1997)

Starring Jodie Foster and based on the novel by Carl Sagan, this is a fictional movie but it does a remarkable job of being realistic in many respects. For instance, it shows that visionary rocket scientists have to endure years of tedium, setbacks, and various forms of resistance that all must be overcome with persistence, creativity, and integrity. If you ever wonder what would happen if we discovered life beyond Earth, this is a must-see movie! Rest assured, if and when it happens, it will be due to the efforts of rocket scientists and aerospace engineers! Maybe you will be one of them....

From the Earth To the Moon (1998)

This is an epic 12-hour miniseries from HBO with Tom Hanks and a production budget of $68M.

October Sky (1999)

Yet another inspirational and true story. This is about Homer Hickam (played by Jake Gyllenhaal) who grew up in a coal town and knew he was destined for a different life. He went on to work at NASA. If you are the first person in your family to talk seriously about college or a highly technical career, you need to watch this movie!

You'll notice that this list ends in 1999. Are there more recent movies that you think deserve to be included in this highly esteemed list? By now you should know what to do: tell me about it with email, LinkedIn, or Twitter!

TIP #6:

MAKE A SHORT TAKE-OFF (STO)
INTO THE WORKPLACE

Here is a tip that might surprise you—there are many ways to get into rocket science that don't require a four-year college degree!

Aerospace companies have a wide range of staff and trades people. There are a lot of jobs that require a two-year degree or a certification. Some examples are Computer Numerical Control (CNC) machine operation, financial analysis, composites fabrication, electronics, nondestructive testing, aircraft or engine maintenance.... These are real and meaningful jobs!

If you are a very hands-on and tactile kind of person, this may be your calling. Whether or not it's your long-term job, the point is that your full-time entry into the workplace can be accomplished in half the time as others may take.

Once you are employed, a four-year degree is still an option. In fact, this is when you can often take advantage of other people's money (OPM) to pay for more education. You see, many (probably most) companies of respectable size that need highly skilled employees have a tuition reimbursement program. If you want to take classes or pursue a degree in a subject area that benefits the company, and you get passing grades, the company will pay for the costs!

Each company's policy needs to be known and understood fully, of course. And one tradeoff is that you will have to be doing your job and going to school at the same time. But consider that you will be earning money with a job in the field you love—and getting more education at no cost to you!

There are famous stories about people who started on the shop floor or mailroom and worked their way up into the executive suite. It wasn't easy or common back in those good ole glory days, and it isn't easy or likely today, either. But it does happen! I worked with a talented and ambitious woman on the V-22 Osprey program (let's call her Roberta) who never got an engineering degree. She started with a business certificate. Then she got a business degree and over the years got other credentials. Eventually she became a Program Manager with many engineers reporting to her.

Here are two more powerful short take-off approaches into the workforce: internships and co-opping. If you are set on pursuing a Bachelors degree, make every decision within your power to choose a school that has a strong and successful internship or co-op program! Hopefully, this is an obvious advantage to you: By the time you graduate, you will already have relevant, practical, and (usually) income-producing work experience!

You can make a list of the pros and cons for taking time away from college to work in the field versus going straight through. But I'll give you my story. As a freshman at Purdue University in Indiana, I applied to the School of Aeronautics & Astronautics for my sophomore year and I also applied for the co-op program. I worked myself really hard freshman year and got accepted to both programs (success!). My first co-op session was the fall semester of my sophomore year, which was at McDonnell Aircraft Company in St. Louis, Missouri (later part of Boeing). It was in the center fuselage structural design group of the F-15 Eagle. It was an exciting, fun, enlightening experience. It some ways it was also scary and humbling, with so many new environments and situations to be in.

By the time I graduated I had FIVE real and fascinating work assignments in FIVE DIFFERENT departments! For me, it confirmed that this was the profession and career that I had to do. Even nicer—it confirmed for my sponsor company that I was worth hiring upon graduation! And it was a very rough economy then, as it is now. (It was at the end of the Cold War.) I later learned that the company made only TWO offers to engineering graduates from Purdue that spring. Both of us were from the co-op program—a female classmate and me. Clearly, the co-op program was a major win-win for everyone. It launched my career off to an early and exciting start.

In short, be creative, resourceful and ambitious to get yourself into industry as soon as possible. What's the worst that could happen? You find out you hate working at a certain job or in a certain company? Maybe rocket science isn't for you? Hello! That's a great lesson and result!! You just learned it a lot sooner and cheaper than if it happened after graduation and with a permanent job!

Job-shadowing, career auditing, free internships—do what you can to make a short take-off into the field of rocket science so you can start your career as soon as possible and start moving forward from there.

Additional Resource:

The University of Texas has done an amazing job listing all US community colleges by state and city in one place: http://www.utexas.edu/world/comcol/state/

TIP #7:

USE FREE AND INFORMAL
EDUCATION RESOURCES

There is a wealth of resources and opportunities for you to gain knowledge, experience, and a competitive edge that can help you enter the world of rocket science. And the best thing is, they can be tailored to your specific needs and interests. But you need to put *yourself* in the pilot's seat!

With YouTube, Vimeo, and other websites, you can have access to tremendous informational videos for free. How about searching there for your favorite rocket science topic instead of watching pet videos? (Or Russian dashcam videos in my case…at least those have lots of physics, I tell myself.)

The Khan Academy is excellent for bite-sized math, physics, and science lessons (plus a whole lot more).

Many universities and other educational institutions are now offering MOOCs: Massively Online Open Courses (pronounced "mooks" – rhymes with kooks). One of the most famous is MIT. Yes, these take time. So decide to have some self-discipline and put your time to productive use!

In terms of informal education, there are camps, competitions, and project-based activities. If you are a student in middle school, high school, or college you have many options. If you are an adult, you may not think these are relevant for you. But guess what—you will learn things too!

Even as a volunteer. And, you are going to meet the other volunteers who are probably (you guessed it) rocket scientists already working in the field. So look for aerospace or aviation-themes camps, competitions, and museums in your area.

These are very specific and dependent on geographic locations, so you will have to do more research yourself. But here are some examples and web search terms to start with (and try adding your city after these keywords):

STEM camp

STEM summer camp

Robotics camp

Robotics club

Women in Aviation

Design build fly competition

EEA Young Eagles

Rocket Challenge

US FIRST

Women in Aerospace

First Lego League

Future City Competition

Team America Rocketry Challenge

AIAA Design Build Fly Competition

Additional Resources:

- The Khan Academy is here: www.khanacademy.org.

- MOOCs from Embry-Riddle University (an aviation focus) are found by starting at their home page here: www.worldwide.erau.edu.

- MOOCs from MIT are located at www.ocw.mit.edu. In particular: AE 16.885 is Aircraft Systems Engineering. This is powerful and exciting stuff!

TIP #8:

KNOW HOW TO APPLY FOR A JOB

First of all, let's question this assumption. Even if you are currently in college or looking for your first job in rocket science, what if an employer or future boss felt the need to pursue you?! What if you got a personal invitation to start working with someone? This is a more powerful and fulfilling position, isn't it? Keep reading and we'll work on this.

Applying for a job at a large aerospace company has changed over the years. Today it's like most large companies in any industry: very process-controlled, website-driven, and structured. There are good intentions and business constraints that have driven us to this state of affairs. But in most ways, it is <u>not</u> for the benefit of you, the job seeker. If you haven't read and really focused on Tip #6 yet, do that first. Now, when you get to a point where there are some jobs (or internships) to apply for, here are some tips for that.

The better company websites have a way for job seekers to create a profile with search preferences. Do that. They should provide a way for you to be emailed when any new jobs are posted that meet your criteria. Use that! Automation is smart here—put those tools to work for you.

When you find a job you want to apply for, please remember these guidelines:

- Your resume needs to be tailored to that particular job! This means to emphasize education or work experience that matches elements of the job description.

- There are automated screening algorithms that filter out resumes. Look for key responsibilities or requirements and use the same words in your resume.

- If you don't use any of the key words from the job description or don't meet the minimum requirements they specify (e.g. "[X] is a requirement" or "[X] is required"), don't bother applying. Your time is better spent on things that have a higher likelihood of positive outcomes. One of the many pieces of wisdom I heard from my early co-op assignments was "Don't confuse activity with accomplishment." I've repeated that to myself—and others—many times since then!

- In many ways, this is a numbers game. It takes volume. Companies are dealing with volume, too. A Human Resources (HR) person told me in 2013 that they get close to 100 resumes for every job posting. Yikes! Thus the next point:

- Think about and include anything in your experience that can help you stand out. (More on that soon!)

- Use specific numbers in your resume, especially when it comes to responsibilities and accomplishments. For instance, how large a budget was your project? How many lines of computer code did you write? What percent increase in improvement did your project produce? These numbers are hard (or dishonest) to make up after the fact. So, get in the habit now of defining and tracking key metrics for projects that you work on so you can have them later. (More fundamentally, they will help you focus on what's important for your projects!)

- Don't wait for a response before applying for other jobs. It's ok to hope—but as one of my best bosses was fond of saying, "Hope is not a plan." Once it's out of your control, get back to doing what IS in your control. This means: continue to look for and apply to other jobs! And make yourself more unique and valuable to future employers (with more on that soon).

So how do personal connections help, if at all? Good question! And they do! Here's how:

Knowing someone in the company with the job opening might get you more information beyond the job description. But be aware and be warned—if you try to talk to the hiring manager outside of the formal process, that is a big no-no and will cause disqualification. It will get the manager in trouble, too. So, the more people you know in a company, the more likely you'll have someone safe to talk to.

There are also some companies or jobs that still rely on word-of-mouth networking. Learning about a job before it gets posted also gives you more time to prepare and position yourself for it. You could have several months to prepare.

The smartest and ultimately most successful approach in this modern era, in my humble opinion, is to make yourself discoverable and compelling to potential employers *before* you even contact them. <u>Without</u> you needing to contact them!

Think about this seriously—each one of us is our own media company now! Whether you want it to be or not, every artifact you post online or appearance you make that is uploaded to the web (even from your friend with the fast photo finger) is one of your media and public relations products. If you want to be an aerospace professional (or professional in any field, for that matter) you need to start thinking about proactive <u>offense</u> and <u>defense</u> when it comes to your media strategy and outputs.

One of my most all-time favorite music concerts was when I was in graduate school at Penn State University. Billy Joel came to town. (I play piano and used his songs to learn piano and the art of songwriting as a teenager, so any concert with Billy Joel is special to me.) This was one of his interactive "conversation" concerts where he would take questions from the audience. Every chance for a new question, I would jump up and down, wave my arms wildly and scream, "Pick me! Pick me!"

That was me—along with a few thousand other people.... And that, my friend, is what it's like to apply for a job at a big aerospace company today. (And no, he never picked me—even though I was in the 3rd row! We even made eye contact.)

Fast forward a few years, and I was still thinking (ok, daydreaming) about an album of original songs that I wanted to create. Some day....

Then I learned about CD Baby, which lets anyone upload, burn, and sell a music CD for under $50. No more permission or approval from a record label or professional music studio required! Home music recording software was inexpensive and plentiful by this point, too.

It took another year of persistence, learning, and many late nights after work, but I finally got twelve songs recorded and packaged into a piece of artwork that I was proud of. Then with some courage and fear, I shipped it out to the world.

Bruce Lee, the martial artist movie star, is one person quoted with the phrase "Courage is not the absence of fear. It is the ability to act in the presence of fear."

What kind of rocket science work (or *artful work*) do you dream of creating? Allow yourself to daydream and visualize it. Shape it into a clear and compelling vision. Then think about all of the platforms or outlets you have available to you today that don't need anyone else's permission or authority to use. You can start learning, advancing, and creating—today!

Maybe it's podcasting. Online videos. A website or blog posts dedicated to the discipline or technology that you are most passionate about. Mobile app development. Seek out the current experts and ask them if you can interview them (after you've learned everything you can from what's available already).

Use this passionate topic and the media channel of your choice to learn, share, and create (then repeat) with the rest of the aerospace community able to benefit from it, from you, and with you.

Oliver Wendell Holmes Jr. said, "most men and women go to their graves with their music still in them." Don't let that be you! How, you may ask? As Seth Godin urges in *Poke The Box*, "Reject the tyranny of the picked. Pick yourself."

Here's a random rocket science example. Suppose you are really fascinated by thermal heat shields (for hypersonic or re-entry vehicles). Suppose you already know how to program mobile apps and know something about game design. Boom! There's a unique combination and offering. (At this point in time it appears to be, at least.)

If you feel stuck doing the work over and over again without any positive results, or still don't know how to get started after checking out the resources below, it's time for Tip #9!

Additional Resources:

- Learn why and how you should be irreplaceable with *Linchpin* by Seth Godin.

- Encore recommendation! A best-selling career guide book that has great exercises to create your own unique path, updated every year, is *What Color is Your Parachute?*, by Richard Bolles.

- Learn the value and thrill of taking initiative and seeing what happens with *Poke The Box* by Seth Godin.

- Want some relaxing instrumental piano music composed by a rocket scientist? Check out *Preludes & Reflections* on cdbaby.com by Brett Hoffstadt (yours truly).

TIP #9:

FIND A MENTOR

What is one thing all of the heroes in epic journeys have in common? A mentor! A guide, a sensei, a Jedi Master....

A mentor can provide priceless advice, coaching, and connections. Mentors tell you things that you wish were taught in school, and maybe are supposed to be, but weren't. A good mentor doesn't just tell you what TO DO, but what NOT to do. And a great mentor helps you reveal your strengths and your blind spots that you aren't able to see for yourself.

Here is something else every epic journey has in common: struggles. Heroes and heroines will hear people say discouraging things like "You can't do that," or "Why bother trying?" Or even worse, "What makes you think *you* could do this?"

One powerful defense against this challenge to your mission is to know (and say), "I know that [my mentor] grew up in a tough situation, in some ways even worse than I did. Today she (or he) is successful doing great things. And they aren't alone."

Have you watched "October Sky" mentioned in Tip #5 yet? Homer Hickam overcame intense pressure from family and community to pursue a career that wasn't in a coal mine. His first mentor was a school teacher who encouraged him to learn and play with things he loved (model rockets).

One of my early experiences that was crucial to get me hooked on aerospace was an after-school program at the nearby NASA Center (then NASA Lewis, now NASA Glenn). I was in the 5th grade. An engineer there stayed late after work one day a week for about five weeks to work with me and a group of kids. We designed, built, and then flew our own airplanes made out of balsa wood, tissue paper, and rubber-band-powered propellers. He encouraged and energized my passion for airplanes in ways that I still am thankful for today. (One way I show thanks is by continuing the circle as a volunteer in classrooms and community events.)

Teachers and community volunteers can be priceless mentors as these two stories show. If you are at the age when you can meet someone like this, keep your eyes and ears open for the opportunity.

If you are older, there are several good books, blog posts, and articles about how to find a mentor. Here are some common suggestions and tips:

- Know what you want ahead of time. What kind of advice or expertise are you looking for?

- Use networking events and LinkedIn to find potential mentors.

- Start with small and specific requests or areas for mentoring. Don't jump into a big proposal or a complete life-coaching experience. Start small and build a relationship over time.

- Reciprocate. Be genuinely interested in asking how you can help your mentor.

- Listen, act on their advice, try to learn from it, then report back to them. Be an active participant in the process.

- Respect their time. Work around their schedule and constraints, and be the first to honor the time constraint when you've reached it.

- Be pleasant, fun, and easy to get along with. A mentor is probably not looking for another "project" or employee, so don't act like one!

- Be reliable!! There is nothing more disrespectful than being a no-show (in my opinion, at least). And if you agree to do something (as in taking some advice or taking some action), do it! Or be open and honest about the root cause of why you didn't. Maybe *that's* the issue you need a mentor to help you with.

You can have more than one mentor. And they will probably change over time and circumstances. But treasure each one and strive to be in a learning mode with each.

May the force of a good mentor be with you!

Additional Resources:

- If you have joined a professional society or club after reading Tip #3, investigate a mentoring program through that organization.

- An excellent online article is from author and blogger Karen Burns from 2010: 13 Tips on Finding a Mentor. Go to money.usnews.com, then search for "tips on finding a mentor."

TIP #10:

APPRECIATE AND WORK ON THE SOFT SKILLS

One of the famous (or infamous) interview questions a person can get is some version of "What's your biggest weakness or flaw?"

I was a graduate student when I first got this question. In a rare and ill-timed burst of self-awareness, I said, "Well, I'm really not much of a people person."

D'oh!! I could see the interviewer scratch me off the list right there—literally! He wrapped it up quickly at that point.

It took me more years than I want to admit, but I eventually realized that people skills and "soft skills" make a tremendous difference in a person's career—for better and for worse.

Hopefully, you won't have as long and hard an effort as I've had to appreciate this and gain some competency. But, hopefully my story will also show that even a clueless, awkward person like me can change for the better. You're about to learn how I did it. (And how I'm still working on it!)

There are some rocket scientists who can be "successful" while also being difficult to work with—or happy working in their own solitary space. They are the rare exception. Our heavily connected and collaborative world is making that option increasingly risky and unsustainable.

Assuming you plan to have some regular interaction with other people, there are <u>three</u> powerful and proven resources that will provide their own list of tips and lessons worth your time and investment. The first is the classic book by Dale Carnegie, *How to Win Friends and Influence People*. It is a must-read! For more intensive and impactful results, the Dale Carnegie organization holds workshops that transform your communication and leadership abilities.

Another proven, fun, and immensely rewarding organization is **Toastmasters International**. Joining a local club will give you the opportunity to <u>learn by doing</u> in a very supportive, time-efficient, and friendly atmosphere. If it weren't for Toastmasters, I'd still be threatening my career with almost every interpersonal interaction. Delivering effective presentations and running efficient meetings are essential tasks for rocket scientists. You'll learn how to do both (and more) with confidence and poise through Toastmasters.

Also worth mentioning here in the non-technical realm—there are some basic attitude and behavioral qualities that you will need to be successful. In no particular order—because they all are important!

- Be reliable and trustworthy.

- Be respectful.

- Be friendly.

- Be humbitious: humble yet ambitious.

In short, be your best, authentic self. You want to give other people good reasons to work with you. And you want to minimize reasons or excuses for them NOT to work with you!

Additional Resources:

- Toastmasters International (www.toastmasters.org) gives you great practice and growth in a supportive, fun, and friendly environment.

- A must-read and classic book of wisdom: *How to Win Friends and Influence People* by Dale Carnegie.

- The Dale Carnegie organization offers training, workshops and more.

READY FOR TAKEOFF

In most books, this chapter would be called the Conclusion. But for you, it's a new beginning.

What is the hardest part of getting into orbit? It's overcoming the initial inertia to clear the launch pad. So much of the fuel is used for the early ascent that engineers go to create lengths to work around it, such as building a floating launch platform that can move to the equator (e.g. Space Launch) or building a carrier aircraft to start the rocket launch from a much higher altitude (e.g. Virgin Galactic).

Similarly, with passenger aircraft, it's the takeoff conditions that dictate the size and thrust of the engines. If it weren't for this phase of flight, the size of the engines could be much smaller. This would reduce the amount of fuel required onboard, which would also reduce the weight of the structure...flying would be much easier and cheaper in many ways.

But we all have to start from where we are. The hardest part is making the decision to start—and then acting on it (which means you *really did* make the decision).

If this is your first step towards a career in rocket science, congratulations! And thank you! We need you to keep moving ahead so you can join us. Visualize the future state that you want to achieve—your own version of a cruising altitude or orbital condition. Then we need you to chart a course and make a plan to keep moving ahead. Take another step forward. Put some energy, action, and momentum into your mission.

Rocket science ain't easy, as they say. But it is possible! It is happening! And in the grand scheme of things, we humans are just getting started. People around the planet (and above it) are learning and doing great things in aerospace and aviation every day. We have all had to overcome various obstacles of ignorance, resistance, doubt, fear, frustration, adversity.... But, we persisted and innovated to keep moving forward toward our goals.

Whatever your personal challenges may be, now or later in your journey, remember to keep your dream and vision alive. You may have to throttle back on a current course. You should expect to plan an alternate route around some unexpected turbulence. Acknowledge and accept right now that "stuff" happens to everyone and life isn't always fair.

Just remember to keep your spirits up, keep your wits about you, and keep an eye on the horizon. Then we'll see you somewhere in the awesome world of rocket science and aerospace.

Take care, and take charge!

ABOUT THE AUTHOR

Brett Hoffstadt is an aerospace engineer (aka rocket scientist), project manager, inventor, music composer, and innovation catalyst. When he thought about what career he wanted to pursue, he thought about the hobbies that he loved most: building and flying radio controlled-airplanes and playing the piano. Being practical, airplanes won. He went on to earn a Bachelors of Science in Aeronautical & Astronautical Engineering (BSAAE) from Purdue University and a Masters of Science in Aerospace Engineering (MSAE) from Penn State University, with considerable industry experience in the meantime.

Many years later, Brett (finally) figured out how to combine his creative energies (previously applied to music) with his analytical and engineering skills. While overseeing development, production, and multidisciplinary engineering efforts on the V-22 Osprey tiltrotor, he activated a new culture and capability of innovation within his large aerospace employer with such additions as 3-D printing, dedicated innovation spaces for employees, and an internal crowd funding platform.

In addition to managing and leading projects of the innovative, complex, and aerospace varieties, Brett delivers presentations and workshops on aerospace/STEM topics and creating innovation within organizations. His website and blog are the sources for further discussions and insights into aerospace career trends, tips, and resources: www.howtobearocketscientist.com

To contact Brett, look for him on LinkedIn, send an email to brett@howtobearocketscientist.com, or connect on Twitter @BrettRocketSci.